Viral
Spiral

A Statistician Analyzes COVID-19 in America

BENJAMIN HALL, PhD

CONTENTS

1
INTRODUCTION: COVID-19 IN AMERICA

I wanted to know.

Was COVID-19 going to kill as many people as experts predicted? Were drastic government actions really what was required to flatten the curve? What should I do to keep myself and my family safe?

As a statistician, I was committed to following coronavirus by the numbers. Enough results are in, and I am ready to share my findings with you.

You probably have questions yourself, like:
- Did my state overreact or underreact?
- Is it safe for me to go out in public?
- Am I putting others at risk if I do?
- Should I wear a mask?

Fortunately, most of these questions can be answered numerically. Each of the 50 states took its own approach. We can compare their results. Comparisons can also be made between coronavirus decisions and risky choices you face every day. This book will reach statistical conclusions with a substantial degree of certainty.

But at the start of the outbreak, things were much less clear.

In an effort to enforce social distancing, a Kentucky doctor shoved apart teenage girls congregating in a park.[1] In a neighboring county an entire church continued to meet in person that same week, defying government restrictions.[2] These actions were on extreme ends of the spectrum, but it was difficult to know who was right about the danger.

The discordant pandemic postures and changing directives stretched to the highest levels of government and academia. Dr. Anthony Fauci emerged as the most recognized American coronavirus expert. On January 26 he stated that COVID-19 was "a very, very low risk to the United States".[3] By March 15, his message had changed to: "If it looks like you're overreacting, you're probably doing the right thing".[4] By the end of March, the public projections for what might happen in terms of American deaths ranged from five thousand[5] to five million.[6]

Experiences differed just as dramatically as the projections. British Prime Minister Boris Johnson and thousands of others contracted the coronavirus and ended up in the ICU. Basketball star Kevin Durant and thousands of others contracted the coronavirus and ended up symptom-free.

I realized just how difficult coronavirus predictions could be when I modeled my own projections in mid-March. In less than a week I had to correct major assumptions based on emerging data. But more data brings more insights and more confidence, and by the end of March the main statistical conclusions of this book were already coming into focus. April brought confirmation that three conclusions are merited:

1. Drastic government restrictions did not significantly alter the course of COVID-19
2. The death rate for COVID-19 was much lower than we were initially told by experts (and much, much lower than the rates you have seen quoted on cable news)
3. The COVID-19 outbreak, like all disease outbreaks, did not grow exponentially anywhere

This book is about my journey to analyze COVID-19 by the numbers, but it is also about *our* story as we have all experienced COVID-19 in America.

2

THE CURVE

Subsequent generations will never be able to comprehend the experience of the week leading up to March 15, 2020. News of a novel and deadly coronavirus had been circulating for weeks, but the story had been about China and then Italy and then a cruise ship in the Pacific Ocean. Americans were merely spectators. Despite a 15% drop in the stock market, Americans largely felt safe as February closed.

Then confirmed cases began to mount, and it became clear containment was not going happen. The virus was here and the only variable we could control was our response.

Three pieces of information crystalized quickly in the public psyche: (1) the coronavirus was deadly, (2) asymptomatic carriers could spread it, and (3) the country would need far more ventilators to support virus victims than we had in capacity. There was reason for the alarm.

The World Health Organization (WHO) announced the death rate for the coronavirus was 3.4%.[7] Plus, infected individuals commonly lacked symptoms, but they could still spread the disease.[8] This translated into an extremely popular and effective social media meme:

One phrase did more than anything else to drive home the need for action: "Flatten the curve." Americans could read the news and see that Italy was facing a health crisis, with intensive care units at capacity. Many patients needed ventilators to survive. Many did not survive.

Epidemiologists quickly estimated what would come next, and a well-publicized estimate was that America would need 1 million ventilators but had fewer than 200,000 in stock.[9] The only viable solution was to flatten the curve. LiveScience.com stated succinctly, "a flatter curve assumes the same number of people get infected, but over a longer period of time."[10]

We took the only actions we could take. Governors began sounding public warnings and issuing restrictions which, at that time, were nothing short of shocking. On March 12, Ohio Governor Mike DeWine banned gatherings of more than 100 people.[11] There were five confirmed cases in his state. The NCAA basketball tournament was scheduled to kick-off soon in Dayton, but the NCAA had already banned fans from the tournament on March 11.[12] On March 15, Ohio became the first state to close restaurants to indoor dining. Eighteen more states followed suit the next day.

The pivot toward fighting the coronavirus felt like whiplash. On Monday, March 9, I had shaken hands with new acquaintances who had flown into Kentucky from Florida specifically for a large gathering. Later that evening I had skated with my kids in a crowded rink. By Sunday, March 15, churches had almost universally cancelled in-person services and the NCAA had cancelled March Madness entirely.

The unthinkable had happened. And yet the models told us the worst was yet to come.

3
THE MODELS

The shift from complacency to fear in America was driven by what we observed. Senator Tom Cotton was the earliest member of Congress to sound the alarm, publicly urging the FDA to expedite testing on January 26. He told reporters his prescience was not due to his scientific credentials but because he "has two eyes" and realized that China's coronavirus problem was severe.[13]

It took six more weeks for average Americans to become alarmed as we read heartbreaking stories out of Italy. A doctor's message was representative of many anecdotes that made their way across the Atlantic Ocean: "It's not a wave. It's a tsunami."[14] Italian hospitals had such a relentless stream of patients that the country imposed a nationwide lockdown.[15] Italy was sending us a warning. Graphics hit social media showing that the US was only ten or eleven days behind Italy's trajectory.

Sufficiently unsettled by our eyes, we turned to scientists for directions. Specifically, we sought scenarios supported by statistics to guide our steps. Academics answered the call, and models began streaming from computers. One conclusion the models had in common: infections, hospitalizations, and deaths were going to get really bad, really soon.

Three projections were especially influential: (1) a report from Imperial College London which was shared with government officials across the developed world, (2) a website model published at CovidActNow.org that was endorsed by a bevy of scholars, and (3) data conclusions released by the White House Coronavirus Task Force. Absent government intervention, all three models agreed that American deaths would top 2 million. The models were also unanimous that most deaths could be averted if aggressive action were adopted, even if the remaining volume of deaths was still very grim.

The Imperial College Model
Though he had 30 co-authors, lead author Neil Ferguson became the face of the British model with the official title "Report 9".[16] It would be fair to call Mr. Ferguson the most influential person of 2020. His model's main focus was Great Britain, but it covered the United States as well and was distributed to the American government during the whirlwind week of March 15.[17] Coronavirus task force leader Dr. Deborah Birx pointed to this model on March 16 as the rationale for implementing the national coronavirus guidelines called "15 Days to Slow the Spread". But this was an optimistic spin on the model's recommendation, which called for 18 months of intense suppression.

Here are highlights of what the model assumed:
- One third of transmission occurred among households, in schools/workplaces, and in the community, respectively
- Two-thirds of cases were symptomatic
- Symptomatic cases were twice as likely as asymptomatic cases to transmit the virus
- 4.4% of infected individuals would be hospitalized, varying dramatically by age
- 0.9% of infected individuals would die, varying dramatically by age

The model simulated five interventions:
1. Case isolation: Symptomatic individuals stay home for 7 days
2. Home quarantine: Symptomatic individuals and their family stay home for 14 days
3. Social distancing of the elderly: Those over age 70 are forced to dramatically decrease contacts
4. Social distancing of everyone: All individuals are forced to dramatically decrease contacts
5. School closures

The model came to the following conclusions:
- Without government intervention, America would have 2.2 million coronavirus deaths
- To avoid hospital overload and catastrophic deaths, all five interventions were necessary
- Even if all five interventions were deployed 70% of the time for more than a year, America would still have 168,000 coronavirus deaths

The situation was dire, but the prescription was obvious. The world was in this together; the authors described the model as "equally applicable to most high-income countries".

The models were in this together, too.

CovidActNow Model
The projection website CovidActNow.org racked up academic endorsements as soon as it was published. Experts from Yale and Georgetown and the Health Departments of Alaska and New York gave their approval. Former CMS administrator Dr. Donald Berwick had served under Barack Obama and been named a Fellow of the Royal College of Physicians in London; he endorsed the simulations as well. This was a model to be taken seriously.

The purpose of the tool was to help leaders "better understand the future spread of COVID." Leaders soon took notice as the model was cited by Michigan governor Gretchen Whitmer, Kentucky governor Andy Beshear[18], Dallas judge Clay Jenkins[19], and the Health Department of Alaska.[20]

The model underwent several later iterations, but the March model included projected deaths by state under four scenarios:
1. No Action
2. 3 Months of Texas-style delay/social distancing
3. 3 Months of California-style "shelter-in-place"
4. 3 Months of Wuhan-style Lockdown

The potential deaths were more dire and the potential impact of interventions more dramatic than the Imperial College model. For example, Florida alone could expect the following within 3 months:
- 432,000 deaths with no action
- 320,000 deaths with a Texas-style delay
- 8,000 deaths with a California-style "shelter-in-place"
- Fewer than 1,000 deaths with a Wuhan-style Lockdown

The clear message from this model was that the previously unthinkable actions from the week of March 15 were wildly insufficient. More action was needed, and soon. The website listed a "point of no-return for intervention to prevent hospital overload" for Florida as falling between March 25 and March 30.

Coronavirus Task Force Model

The White House Coronavirus Task Force released the results of their models on March 31. The details were sparse but the essence was distilled in a poster presented by Dr. Deborah Birx while wearing one of her trademark bold scarves.[21] No intervention meant 1.5 million to 2.2 million American deaths, while the scenario "with intervention" meant only 100,000 to 240,000 fatalities. This range was actually higher than she had indicated the day before when she had cited 100,000 to 200,000 deaths, with Dr. Anthony Fauci echoing the same range. Importantly, Dr. Birx added that keeping deaths below 200,000 would only happen if "we do things almost perfectly" with "100 percent of Americans doing precisely what is required".[22]

These three models largely set in place the course of events that followed, with the Imperial College model asserting an especially large influence. (Dr. Birx alluded to information from a model developed in Britain in her remarks.) Soon Americans experienced an enlarging patchwork of restrictions: business closures, stay-at-home orders, and even restrictions on where you could drive. And there was no guarantee that it would work.

Full-blown academic models were not the only source for coronavirus metrics and projections. Politicians, medical experts, and data scientists across the country went on record, and only a very few wagered anything less pessimistic. Here is a quick run-down:

March 10
- The headline of an article by former CDC director Tom Frieden seemed alarming: "Could Coronavirus Kill a Million Americans?" But the text was scarier still. His worst-case scenario was 1,635,000 deaths in the United States.[23]

March 13
- The media published CDC projections that 200,000 to 1.7 million Americans could die from coronavirus within a year.[24]

March 15
- Gretchen LaSalle, MD blogged that we could see "over 5 million deaths in the United States alone."

March 19
- Tomas Pueyo (CovidActNow creator) shared a graphic on Medium.com with the title "Infections and Deaths If We Do Nothing in the US" with a text box overlaying the graphic with the conclusion "Total Dead:

> 10,000,000". To bolster its credibility, the article added endorsements by more than two dozen experts and thinkers representing Oxford, Stanford, Rutgers, and more. The article quickly amassed more than 10 million views.[25]

• California Governor Gavin Newsom projected 56% of Californians would be infected within eight weeks as he issued a statewide shelter-in-place order.[26]

March 25

• New York Governor Andrew Cuomo tweeted: "Our single greatest challenge is ventilators. We need 30,000 ventilators. We have 11,000."

• Dr. Peter Angelos with the University of Chicago Medical Center pitched a lottery system to determine which patients would receive a ventilator as the state of Illinois worried that ventilator demand would double supply by April 6.[27]

• Science beat journalist Ed Yong relayed that "even if social-distancing measures can reduce infection rates by 95 percent, 960,000 Americans will still need intensive care."[28]

March 26

• UCLA's Joseph Ladapo, MD wrote: "Short of a miracle, expect to see a tragedy unlike anything we've seen in generations. Heartbreakingly, people you know will die." He was skeptical that government intervention could make things any less bleak, adding: "It's too late for any other outcome."[29]

• Johns Hopkins' Tom Inglesby wrote: "Anyone advising the end of massive social distancing now needs to fully understand what the country will look like if we do that. COVID-19 would spread widely, rapidly, terribly — and could kill millions in the year ahead."[30]

March 27

• Ohio Department of Health's Dr. Amy Acton said the state had only one-third of the hospital capacity it would need for peak demand in May.[31]

March 28

• University of Pennsylvania's Ezekiel Emanuel, MD wrote that if we did not implement and maintain a national lockdown "100 million Americans will have Covid-19 by early May. If 1 percent of those infected die, there would still be a million deaths."[32]

On March 1, America was largely business as usual. By April 1, these expert warnings were being widely heeded as 30 states banned the operations of non-essential business and 29 states had formally ordered citizens to stay at home.

4
CORONAVIRUS IN CONTEXT

You now obviously know the worst did not come to pass. Any deaths are tragic, but compared to what could have been, the result was a collective sigh of relief. There were hot spots, to be sure. Yet vaccine expert Seth Berkley had predicted that "The U.S. may end up with the worst outbreak in the industrialized world."[33] Instead, the United States has had lower per capita coronavirus deaths than Belgium, Spain, Italy, France, Britain, the Netherlands, Sweden, and Switzerland.[34]

The Institute for Health Metrics and Evaluation (IHME) plotted the COVID-19 resource demand peak occurring on April 19. At that point, the need for invasive ventilators reached just under 17,000. This nationwide demand was far below the 30,000 need Governor Cuomo had projected for the state of New York alone!

Throughout March, medical schools announced early graduations for medical students to care for anticipated patients,[35] and the state of Illinois allowed doctors with expired licenses to begin practicing again.[36] Yet April brought layoffs and pay-cuts for thousands of clinicians as demand for healthcare dried up and the need for coronavirus treatment fell far short of expectations.[37] An army field hospital set up in Seattle was dismantled without treating a single patient.[38]

To be clear, no one should confuse COVID-19 as a mild disease. Many of those infected have suffered from severe cough, pneumonia, loss of smell/taste, and even seizures.[39] It is dangerous and deadly. Sadly, it has not been America's worst pathogen predator.

Adjusted for population growth, the 1957 and 1968 flus killed the equivalent of 220,000 and 165,000 Americans, respectively.[40] The "Spanish" flu killed 195,000 Americans in October 1918 alone.[41] The CDC estimated the winter of 2017-2018 saw somewhere between 46,000 and 91,000 flu deaths.[42] (The deaths in a flu season are not all due to influenza.[43] "Flu season" deaths are caused by a mix of wintry pathogens that are often grouped together as "influenza-like illness".[44]) As of May 4, the IHME projects we are on course for 134,000 coronavirus deaths between February 4 and August 4, 2020.[45]

How did America avoid a catastrophe that seemed inevitable? If we had locked down more aggressively could there have been even fewer deaths? Those are questions we will tackle head on. But first we detour through a surprising Nordic country that approached COVID-19 like no one else in the world.

5

SWEDEN, THE OUTLIER

In his inaugural lecture upon being inducted into the Royal Swedish Academy of War Sciences in 2005, Anders Tegnell gave a lecture in which he compared pandemics to war, concluding that in recent years pandemics had in fact eclipsed war as a dominant threat to humanity.[46] So perhaps it should not have been surprising that Tegnell led Sweden's aggressive vaccination efforts against the new swine flu pathogen in 2010.

Authorities convinced more than 5 million Swedes (roughly half of the population) to receive shots of GlaxoSmithKline's vaccine. The benefit of this government intervention seems negligible in retrospect, as the swine flu strain ebbed fairly rapidly worldwide. But the costs were substantial, with more than 500 children suffering from a narcoleptic side-effect as a result.[47]

Tegnell, now the State Epidemiologist of Sweden, has never acknowledged regret about his swine flu vaccination campaign. Yet his decisions must have weighed heavily on his mind for a decade. By 2020 he had transformed from battling pandemics on war footing to tackling COVID-19 with an approach scorned by his peers as excessively relaxed.

Most of the world, and certainly most of Europe had locked down by the end of March. Restrictions in these countries went beyond school and business closures to severely limit physical movement. Italy banned vending machines and put a halt to outdoor physical activity, even individually.[48] France implemented an 8pm curfew, set up police checkpoints, and threatened fines to violators who left their homes.[49] Finland put an entire region of their country into quarantine[50], and Denmark announced a lockdown before it even had its first death.[51]

Meanwhile life largely continued as usual for many Swedes. Young children never had school cancelled and adults maintained access to gyms and restaurants. Skiers who wanted to enjoy the best snow in decades were permitted to do so even into April.[52] "Sweden has gone for mostly voluntary measures," Tegnall explained.[53]

Sweden was not ignoring the issue of COVID-19. Authorities publicly emphasized hand-washing and remote work, and restricted gathering sizes to 50, banned visits to nursing homes, told those over 70 to stay home, and cancelled in-person classes for teenage students.[54] As much as the emphasis was on voluntary measures, it was also on targeted measures. "We can't legislate and ban everything," argued Prime Minister Stefan Lofven.[55]

The scientific community, both inside and outside of Sweden, met the relaxed approach with unbelief. And criticism. "They are leading us to catastrophe," lamented Cecilia Soderberg-Naucler of the Karolinska Institute.[56] "Close down as much as possible," urged Uppsala University's Bjorn Olsen.[57] The approach was "playing Russian roulette with the Swedish population" claimed Lund University's Marcus Carlsson.[58] A group of 2,300 academics wrote an open letter advocating tougher restrictions.

Coronavirus began spreading within Sweden by at least late February, so Dr. Tegnell's experiment has been running for a while. Who was right? We can answer with some statistical comparisons.

As background, the variance in COVID-19 impact by country has been enormous. Deaths-per-million in hard-hit Belgium are 17 times higher than Finland. Pick any two western European countries at random and you should expect one of them to have double or triple the death rate as the other, due simply to chance. Comparisons with a small sample size are unreliable.

Common comparators for Sweden are the Nordic countries of Finland, Norway, and Denmark, all of which have seen much lower deaths-per-million. Sweden's death rate is triple that of Denmark, but Denmark's rate is double that of Finland and Norway. If the elevated rate in Sweden is due to its lax response, what caused Denmark's rate to be so high? Statistically, these comparisons fall apart because a sample size of four is just not enough.

We can perform a better comparison for Sweden if we expand to consider all western Europe with a total of 15 high-income, high-population countries. Here Sweden falls near the middle of the pack (Table 1). In order to attribute

Swedish deaths to the lax policy, we need to explain the general western Europe variation in a way that shows a Swedish excess.

Table 1: Western Europe data as of 5/1/20

Country	Deaths per Million	Median Age	Population per SqKm	Avg Temp (°C)
Belgium	665	41.6	383	9.55
Spain	531	43.9	94	13.30
Italy	467	46.5	206	13.45
UK	405	40.6	281	8.45
France	377	41.7	119	10.70
Netherlands	286	42.8	508	9.25
Sweden	263	41.1	25	2.10
Ireland	256	37.8	72	9.30
Switzerland	203	42.7	219	5.50
Portugal	99	44.6	111	15.15
Denmark	79	42.0	137	7.50
Germany	80	47.8	240	8.50
Austria	65	44.5	109	6.35
Norway	39	39.5	15	1.50
Finland	39	42.8	18	1.70

What could explain the volatility in results? Older ages, higher population density, and a cooler climate are reasonable hypotheses. We can apply the statistical technique of regression analysis, which allows us to control for the effects of multiple variables by finding the relationship that best fits the data.

Results indicate that older ages and a cooler climate fail as hypotheses for explaining country-level results. Controlling for population density still leaves Belgium, Spain, Italy, and France with more excess deaths than Sweden (Figure 1 – further above the line is worse). Sweden has not had the best results in western Europe, but they are far from the worst. Statistically, we cannot say whether Sweden's lax policy resulted in more deaths.

Figure 1: Western Europe COVID-19 Deaths as of 5/1/20

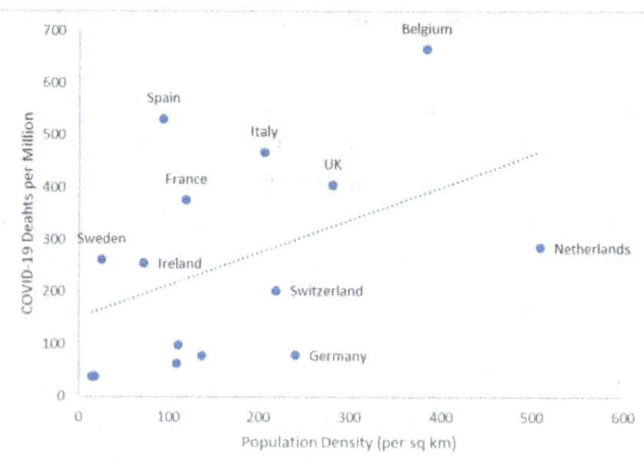

Sweden does provide strong evidence that the alarming expert models were fundamentally wrong. The Imperial College report predicted 255,000 deaths (an off-the-chart 3500 deaths per million) if Great Britain deployed Sweden's strategy of isolating the sick with social distancing for just the elderly, and specifically noted the projections were valid for high-income countries like western Europe. By that measure, Sweden was a grand success and Anders Tegnell should be proud.

While there are only 15 data points in western Europe, there are 50 states which each built its own coronavirus strategy in America. Years ago, Supreme Court Justice Louis Brandeis recognized the opportunity provided by state-level autonomy when he wrote an opinion that "a single courageous state may, if its citizens choose, serve as a laboratory, and try novel experiments without risk to the rest of the country." In 2020 we have seen the states become real-time "laboratories of democracy" with their coronavirus responses, and they offer fruitful ground for assessing strategy effectiveness.

6

THE RESULTS FROM THE STATES ARE IN

When Ohio Governor Mike DeWine issued the first statewide restaurant closures in America, he was in the vanguard of the COVID-19 fight. He was also a Republican. When Virginia Governor Ralph Northam, a medical doctor, was one of the last and most reluctant governors to close restaurants he was described as "steadfast in his limited restrictions".[59] He was also a Democrat.

While Republicans, the party known for more limited government, tended to be less forceful than Democrats overall in implementing restrictions, the fight against coronavirus was not a purely political issue. The COVID-19 strategy was different in every state, with emergency orders ranging from common sense to creative to crazy. Importantly, the many degrees of aggressiveness allow for a robust analysis of the effectiveness of coronavirus government intervention.

The three broad orders which were most impactful to the daily lives of citizens were the termination of indoor dining, the closure of non-essential businesses, and the issuance of stay-at-home orders. The definitional details of these directives differed from state to state, but the timing can provide an especially useful proxy for the overall posture each governor deployed.

The aggressive Governor Mike DeWine issued a stay-at-home order only two weeks after the first diagnosed case in Ohio. The more relaxed Governor Doug Ducey of Arizona issued a stay-at-home order a week later than Ohio, which was 64 days after the first case had been diagnosed in his state. Oklahoma Governor Kevin Stitt, who initially encouraged his citizens to continue patronizing restaurants, was one of a dozen governors who never formally commanded their citizens to stay at home. The starkly different approaches are evident when considering that Florida beaches remained open for spring break, while California's beaches were closed.[60]

As with western Europe, the variation in state-by-state COVID-19 burden has been vast, ranging from over 1,200 deaths per million in New York to only 11 deaths per million in Hawaii. Pick any two states at random and you should expect one of them to have roughly triple the death rate as the other, which is enormous volatility. What drove these differences?

Regression analysis indicates that population density is a significant driver of higher death rates and cooler climate perhaps a small contributor while age is not a state-level factor (though it is certainly an individual-level factor). Urban areas in the northern states endured the brunt of COVID-19's lethal punch. Critically, how rapidly a state responded to COVID-19 with severe social distancing restrictions had no impact on their ensuing death rate.

Consider the two graphs below relating to initial business closures. In the first graphic (Figure 2) we see the raw relationship between response speed (based on initial business closure) and deaths per million. In fact, there is no clear relationship at all. New York's tragic experience came despite a middle-of-the-road response time. The handful of hardest hit states cannot be blamed for hesitation, as they were among the fast actors at the left of the graph. This view does not control for the influence of population density and climate.

Figure 2: American COVID-19 deaths as of May 1 by time until initial business closure

The second graphic (Figure 3) compares actual death rate versus expected from the regression model due to each state's specific features. Even controlling for high population density and cooler climate, New York's excess deaths are more than 900 per million (off the chart). Louisiana's lower population density and warmer climate mean that Louisiana's adjusted performance is even worse than their raw results – second only to New York.

Rhode Island has had a relatively high death rate (Figure 2) but has the best performance in the nation after adjusting for high density and cool climate (Figure 3). Arizona's lax approach did not result in high deaths, landing about where we would have expected.

The big takeaway is the highly symmetric pattern about the horizontal axis in Figure 3. The flat regression (dotted) line shows there was no benefit from rapidly introducing business closures. (This is not to say that individual citizens did not contribute to mitigating COVID-19. More on that in a later chapter.)

Figure 3: Excess COVID-19 deaths by days until initial business closure

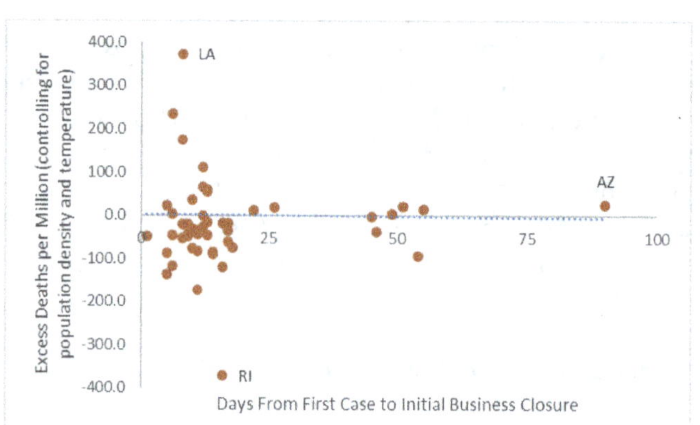

Just as rapid initial business closures did not result in lives saved from COVID-19, neither did stay-at-home orders (Figure 4) nor did comprehensive "non-essential" business closures (Appendix). The graphic for stay-at-home orders contains an interesting story for Massachusetts. Governor Charlie Baker chose the most relaxed approach among the hard-hit northeast states. He never issued a stay-at-home order, while his state was the fourth hardest hit state in the nation. However, controlling for density and climate implies that Massachusetts had fewer deaths than the regression model expected.

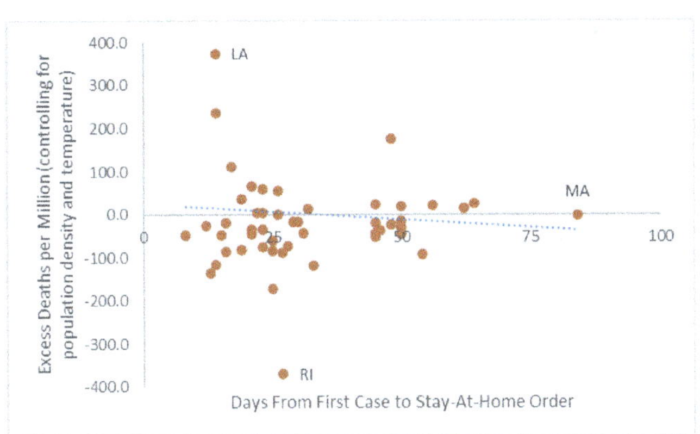

Figure 4: Excess COVID-19 deaths by days until stay-at-home order

The raw data indicates relaxed states outperformed aggressive states but is almost certainly naïve by ignoring density and climate. Controlling for density and climate still leaves the least aggressive states (whether measured by initial closures, extensive closures, or stay-at-home orders) slightly better off than aggressive states. If the slope of the regression (dotted) line in Figure 4 had a very steep upward slope, it would signal such actions paid off. The downward slope instead implies the restrictions were unhelpful.

In summary, statistical analyses of western Europe and the United States lead to the same conclusion: intense government restrictions did not defeat COVID-19. The expert models wrongly forecast that government suppression was a necessary pandemic response.

These models were also wrong in their initial premise that unmitigated COVID-19 would kill millions of Americans, and largely for two reasons: the assumption of a high death rate and the assumption of sustained exponential growth. We will analyze these errant model inputs, but first it is worthwhile to look at the state COVID-19 responses in a bit more detail.

The Opposite of Restrictions
While the unprecedented COVID-19 restrictions upended the lives of most Americans, states also rapidly relaxed laws to grant more liberty in the face of the pandemic. Most of these relaxed regulations targeted healthcare providers, though businesses and individuals were given new options as well.

When the pandemic arrived in America, administrators immediately compared expected hospitalizations to bed capacity and sounded the alarm.

The 1974 Federal Health Planning Resources Development Act required states to tightly regulate the construction and expansion of healthcare facilities. Certificate-of-need regulations ensued and have largely remained on the books even as the federal requirement lapsed.

Most states require government permission for hospitals to add more beds.[61] The idea behind these laws is that hospitals, left to their own devices, will build more capacity than is needed and then have their physicians fill that capacity with referrals that lack necessity. As hospitals sought to expand capacity for coronavirus patients, red tape had to be slashed. [62]

The field hospital in Central Park was only able to be set up because Governor Andrew Cuomo suspended New York's certificate-of-need requirement. Many states similarly waived certificate-of-need requirements to allow hospitals to prepare.

Restrictions on physical capacity were the tip of the iceberg in terms of regulations that could have impeded the coronavirus response. New York responded by eliminating red-tape regulations that consume the valuable time of clinicians. For example, providers were no longer required to follow specific government screening requirements for new patients. Government-required home care visits to certain patients were suspended, as was a government formula used to determine which facility was appropriate for a given patient. The Commissioner of Health was even allowed to suspend energy conservation codes. A legal order was necessary to allow individuals not licensed as lab technicians to perform COVID-19 tests.[63]

Access to telehealth (via webcam or phone) medical care was expanded across the nation for many patients by a simple relaxation of Medicare payment restrictions.[64] Kentucky, Mississippi, New Jersey, and Wisconsin temporarily broadened clinical authority to nurse practitioners by relaxing regulations due to the emergency. California removed the limit on the number of nurse practitioners a doctor could supervise, but otherwise kept their legal limitations in place.[65]

Outside of healthcare, as governors closed restaurants to indoor dining, an immediate secondary problem surfaced. Alcohol sales are generally forbidden on a to-go or curbside basis. State after state quickly eliminated these alcohol restrictions. As the pandemic regulations hit, these relaxed liquor rules led to alcohol being included on 15% of restaurant takeout orders nationwide.[66]

In one of the most creative COVID-19 responses, the state of Maine suspended the requirement that anglers have a fishing license. They wanted to encourage people to social distance in the great outdoors.[67]

Worst Case Regulation
I have made the statistical case that mandated business closures and stay-at-home orders were unnecessary. They also created the need for governments to get far into the weeds on what these closures and orders meant. The result was some poorly reasoned edicts.

If America were to close big-box grocery stores, food access would far eclipse disease as a survival concern. Walmart was always going to remain open as an "essential" business. Michigan Governor Gretchen Whitmer, however, took notice that just because a store was essential did not mean that all items for sale were "essential". She proceeded to ban the sale of "non-essential" items within "essential" stores and ended up in a confused discussion about whether infant car seats were "essential".[68] It is difficult to understand what Governor Whitmer hoped to accomplish with this directive.

Authorities went just as far off the track in handling outdoor recreation. The national guidance was to "practice social distancing", which seems more likely to happen outside than inside. Nevertheless, many parks across the nation were closed. In Louisville, Kentucky, parks remained open, but the parking lots were closed. There seemed to be a shortage of logic in Louisville, as the mayor also banned drive-in church services. "On Sunday morning there would be hundreds of thousands of people driving around our community", he said in justification.[69] (There is no evidence you can catch any virus from driving in your own car.)

In a similar effort to discourage outdoor activity, the city of San Clemente, California filled a skatepark with sand[70], even though skating is not inherently a close-contact activity. Perhaps the worst example was Washington state, which allowed anglers to be in their boat, but not to fish.[71]

The idea to relax alcohol laws came in part from the experience of Pennsylvania where alcohol sales were cancelled statewide. Predictably, the day before the closures took effect, crowds assembled to buy liquor in violation of social distancing guidelines. When Pennsylvanians began crowding West Virginia liquor stores, West Virginia banned the sale of alcohol to Pennsylvanians.[72]

The noble effort to fight COVID-19 had deteriorated in many geographies into barriers and squabbling totally disconnected from pandemic prevention.

Sanity Check

In the analysis above we saw that there is no correlation between various statewide government restrictions and COVID-19 deaths. However, other authors have elsewhere claimed the data does show restrictions are successful. While it is not feasible to refute every such claim one-by-one, we can walk through a couple of flawed arguments that have been made in support of government restrictions' ability to fight virus spread.

To justify restrictions related to coronavirus, a Philadelphia versus St. Louis comparison from 1918 was frequently referenced.[73] When the "Spanish flu" hit the US that year, St. Louis closed schools, libraries and churches only 2 days after their first case, while Philadelphia waited for two weeks and a had a full-blown parade before taking action. Philadelphia had twice as many deaths and some researchers credited St. Louis's restrictions for lives saved.[74] A headline summarized this conclusion succinctly: "A tale of two cities: why social distancing works."[75]

As was the case with comparing Sweden to only three comparators, a Philadelphia-St. Louis comparison lacks a credible sample size. Soldiers brought Spanish flu home from World War I.[76] An alternative theory that Philadelphia was harder hit because it was an entry port for soldiers has just as much support from the two-city comparison. In fact, a study including data on 43 American cities reveals little correlation between public health response time and deaths. Grand Rapids, Michigan, and St. Paul, Minnesota, responded very slowly and fared as well or better than St. Louis. Many cities responded as quickly as St. Louis and yet were hit hard.[77]

Another strategy that provides "proof" of the effectiveness of social distancing is simply to assume the government restrictions are what kept us from the panicked expectations. Such claims are easily refuted by common sense: the news has been just as good in states that were less restrictive.

An example is researchers who predict that Kentucky would have had 44,000 COVID-19 cases due to exponential growth by April 25 without restrictions, when less than 4,000 were observed. They conclude that Kentucky "prevented" 40,000 cases.[78] But no state the researchers examined reached the equivalent of 44,000 cases — even those that were much less restrictive than Kentucky, like Arkansas and Tennessee. The fact is that there were never going to be 44,000 cases in Kentucky by late April. We must stop assuming the models were right and start investigating why they were wrong. That is the topic of the next chapter.

7

THE CORONAVIRUS DEATH RATE

The models were wrong about the ability of government to change the course of COVID-19, but they were also wrong in terms of raw projections. Why did they predict that millions of deaths were going to occur? Two reasons: they assumed a high death rate and they assumed that the disease would grow exponentially. They were wrong on both counts.

The WHO's February Joint Mission report on the China outbreak reported a crude fatality rate of 3.8%. By March 3 the WHO's head Tedros Ghebreyesus announced a fatality rate of 3.4% adding, "by comparison, seasonal flu generally kills far fewer than 1% of those infected".[79] These statements from the WHO were misleading at best, as epidemiologists have long known that initial death rates are artificially high due to low testing rates.[80]

The Imperial College Model and CovidActNow models knew that these death rates of more than 3% were wrong, and instead used rates closer to 1%, which is still a scary 10 times greater than seasonal flu. The worst and least accurate death rate assumption belongs to researchers at the University of Kentucky who assumed a 5% death rate.[81] The high death rates comprised an assumption that I knew was wrong as soon as I dug into the details.

A tragic but enlightening natural experiment occurred when more than 3,000 passengers were stranded aboard the *Diamond Princess* cruise ship in February due to a COVID-19 eruption. Genomic researchers believe the cruise ship outbreak was introduced by a passenger from Washington state.[82]

23

In the close quarters, at least 700 passengers were quickly infected. Importantly, most passengers were tested, meaning the cruise ship contained the highest per capita testing in the world at that time. Since an unknown but large percentage of infections were asymptomatic, the cruise ship instantly became the best place to study COVID-19's lethality.[83]

The raw death rate on the *Diamond Princess* was 1.1% at the end of February. Statisticians knew this was higher than the death rate would be for the general population because the cruise ship was largely comprised of elderly (that is, vulnerable) vacationers. Some academic researchers used this data to true-up the story in China. They figured the true Chinese death rate was 0.5%, about half of what the dire models assumed. The lower death rate was driven by a large volume of asymptomatic cases. Roughly half of those who tested positive were showing no symptoms.[84]

More recent studies have indicated that the proportion of asymptomatic cases may be much higher than 50%. To find the asymptomatic proportion, many asymptomatic individuals need to be tested. (Symptomatic individuals are already tested at a high rate). Broader testing has now been accomplished in several geographies and situations, and each time asymptomatic cases have been the majority, which in turn implies a low COVID-19 death rate:

• Stanford University researchers tested 3,330 individuals in Santa Clara County, California. These individuals were recruited via Facebook ads to represent the entire population as much as possible. They found 50 positives for a 1.5% infection rate. Since only 0.05% of Santa Clara residents had previously tested positive, this represented a 30-fold increase in cases, implying most infected individuals are asymptomatic.[85]

• Columbia University's Irving Medical Center tested 214 pregnant women and found a surprising 33 tested positive for COVID-19. Only 4 of the 33 had coronavirus symptoms.[86] "It is important to recognize that there are a significant number of women (and likely others) who are in the community and asymptomatic," said Dena Goffman, MD, one of the co-authors.[87]

• The state of Ohio deployed widespread testing in prisons in mid-April. They found 1,828 confirmed cases in the Marion County Correctional Institution, which was reportedly 73% of the inmate population. The death rate at the time of the report was 0%.[88] There have since been a few deaths and more positive cases, but the death rate remains far below 1%.[89]

• A company in Iceland offered free testing, and Iceland quickly topped the list of countries in terms of testing per capita. They found a COVID-19 prevalence eight times greater than the official rate, implying a high rate of asymptomatic cases.[90]

While widespread testing has been critical to confirming the low COVID-19 death rate, there was strong early evidence that most cases were asymptomatic. Careful observation and logic led me to this conclusion early in the pandemic.

Consider the age distribution of confirmed cases in the United States as of March 16 from the CDC (Table 2).[91] Logically we should expect age groups to receive roughly equal exposure to the coronavirus. If anything, we might expect natural social distancing to increase with age, leaving the young more exposed. Indeed, the scientific literature supports the expectation that the young are attacked by the coronavirus at the same rates as the elderly.[92]

However, the age distribution of positive COVID-19 cases skews sharply towards the elderly. Logically there must be asymptomatic cases that are not getting tested (Most people think: why get tested if you have no symptoms?).

Table 2: Age distribution of American COVID-19 cases as of March 16

Age Group	COVID-19 Cases	% of Cases	% of US Population	Projected Infections	Projected Unconfirmed Cases	Unconfirmed Rate
0-19	123	5%	24%	1,145	1,022	89%
20-54	1,134	46%	47%	2,241	1,107	49%
55-64	429	18%	13%	620	191	31%
65+	763	31%	16%	763	0	0%
	2,449	100%	100%	4,769	2,320	49%

To arrive at a volume of projected infections consistent with the population distribution we need to project additional unconfirmed cases. At a minimum this requires nearly 50% of cases to be unconfirmed. The unconfirmed cases would mostly be asymptomatic, and certainly not severe. Again, those without symptoms are unlikely to get tested.

Further, this projection assumes that 100% of cases among the 65+ population are confirmed. Correcting for unconfirmed cases among the elderly implies that two-thirds of cases are likely to be unconfirmed and mild.

This exercise reveals that there were likely to be at least twice as many unconfirmed cases as confirmed cases, which in turn implies a much lower death rate than was being reported. Data from Italy, Spain, and France lead to a similar conclusion. As noted previously, do not mistake COVID-19 as a mild disease, but also know that it kills much fewer than 1% of its victims.

Diseases Do Not Grow Exponentially
The Imperial College model projected 2,200,000 COVID-19 deaths in the United States without government intervention and assumed a 0.9% fatality rate. Simple math implies 74% of Americans must become infected for this to happen. The CovidActNow model similarly projected >70% of the population would be infected very quickly without government restrictions.

The argument for such rapid growth has been simply to assume that COVID-19 will grow exponentially if left unchecked. On April 28, the CovidActNow website specifically asserted that cases were growing exponentially in 13 states.[93] A widely shared article in the *Washington Post* stated: "Without any measures to slow it down, COVID-19 will continue to spread exponentially for months."[94]

Anyone who has passed high school algebra can verify that exponential growth would result in nearly all Americans being infected with the coronavirus within a matter of weeks or months. Exponential disease growth is scary.

Simple models often assume that diseases grow exponentially because this is convenient. However, long before COVID-19 mathematicians were critiquing such modeling as inappropriately simplistic.

Public health researchers wrote an article in 2016 noting that "previous work has shown that sub-exponential growth dynamics was a common phenomenon across a range of pathogens, as illustrated by empirical data on the first 3-5 generations of epidemics of influenza, Ebola, foot-and-mouth disease, HIV/AIDS, plague, measles, and smallpox." They concluded that "predictions of final epidemic size based on models that assume early exponential growth will tend to overestimate epidemic size".[95] A 2016 study in the journal *Epidemics* of 20 outbreaks (including influenza, Ebola, smallpox, plague, AIDS, among others) found that all 20 had sub-exponential growth in the early epidemic phase.[96]

Why did experts expect exponential growth when data on other diseases consistently indicates diseases do not grow exponentially? To be direct, this was statistical malpractice.

There are at least a couple of reasons why disease growth is not exponential. The first is inherent differences in the susceptibility and infectivity across the population. In other words, people are different.

As an illustration, consider the case of so-called "viral" videos. You have probably had the experience of stumbling upon a hilarious or fascinating video, only to discover that many of your friends saw it months or years ago. How could you have been so out of touch?

Social media sensations are created when shares lead to more shares which lead to more shares. But the earliest shares are more likely to come from individuals who share more! Over time, the sharing rate declines because the most excitable people got in on the fun first, as they always do.

If you missed out in the early phases of the "viral" growth, the spread may take a while to reach you, if it does at all. This pattern of rapid initial growth which slows and eventually stalls is sub-exponential growth.

The human population has wide variations in frequency and type of contacts, degree of hygiene, and past disease history. These differences interact with disease spread in a way that creates sub-exponential growth.

Another important driver of sub-exponential growth is behavior changes. Rational individuals do not need the government to mandate social distancing when a dangerous disease appears, as we saw with COVID-19.

Data from the reservation service Open Table demonstrated that restaurant traffic collapsed even before indoor dining was officially banned. In Atlanta, dining was down 90% by the time the city's ban took effect.[97] In lax Sweden, venues such as the Abba Museum closed of their own volition.[98] Recall also that the NCAA had cancelled March Madness before governments implemented restrictions.

Once danger is sensed, people naturally take action to eliminate their riskiest behaviors. Importantly, individuals automatically perform situational cost-benefit analysis on their activities. Information about a new pathogen changes their calculations, which decelerates disease growth.

We now know that intense government interventions, such as lockdowns, have no incremental benefit beyond the wisdom of the crowds, but we should have suspected this was true before we tried it.

8

ARE ASYMPTOMATIC CARRIERS DANGEROUS?

The rationale behind the most restrictive COVID-19 emergency orders hinged on the assumption that asymptomatic carriers were a severe problem. "Stay at home, it could save lives," was the advice that accompanied the commands. Here we will crunch some numbers to determine just how dangerous asymptomatic carriers could be. This will be technical but brief; skip to the conclusions if necessary.

How Dangerous is Asymptomatic Spread?
If you are not showing any symptoms, the probability of having the coronavirus depends simply on the general prevalence of the virus in the population and the rate of asymptomatic presentation among all the infected. Bayes' Theorem is a statistical tool that allows us to calculate conditional probabilities. If we call the prevalence "P" and the asymptomatic rate "A" then an application of Bayes' Theorem implies the probability of having the virus given no symptoms is $(A)(P)/[(A)(P)+(1-P)]$. Reasonable choices for A and P would have been 50% and 1% early in the pandemic, and this results in an infection probability of 0.5% for someone with no symptoms.

If we further assume the death rate for the coronavirus is 0.5% and that those infected will with certainty pass it along to another person if they do not isolate, then the probability of a non-isolating asymptomatic individual causing a death is 0.003%.

<u>Conclusion</u>: Probability of asymptomatic spread resulting in death is roughly 1 in 40,000.

How does the risk of deadly asymptomatic spread compare to other decisions we commonly face?

How Dangerous is Not Getting a Flu Shot?
Consider the case of skipping the flu shot. In a typical year we might expect 50% of Americans to get a flu vaccine which is itself 60% effective. Overall, 5% of the total population might get the flu. This implies that the unvaccinated population has a 1 in 14 chance of getting the flu, while the vaccinated population only has a 1 in 35 chance.

Since the flu death rate is 0.1% and again assuming an infected individual will pass it along with certainty, the increased probability of causing a death resulting from neglecting the flu shot is 0.004%

Conclusion: Probability of flu spread from skipping flu shot resulting in a death is roughly 1 in 23,000.

You are putting others at far more risk if you neglect the flu shot than by venturing outdoors during a coronavirus pandemic. Yet millions of Americans neglect the flu shot for long stretches of time.

How Dangerous is Avoiding Radon Mitigation?
Consider the case of indoor radon, which causes 20,000 deaths in the United States annually according to the CDC.[99] The Environmental Protection Agency indicates the lifetime risk of lung cancer death from living in a home with a radon level of 2-4 pCi/L is 0.2-0.5% higher than if radon is mitigated to a level of 1.25.[100] Millions of American homes remain dangerously untested and unmitigated, leaving family members at risk.[101]

Conclusion: Probability of skipping radon mitigation resulting in a death is roughly 1 in 400. Mitigating high radon in your home is roughly 100 times more impactful than isolating during a coronavirus pandemic.

Suppose I was forced to choose among three situations: (1) asymptomatic people interact as normal during the COVID-19 pandemic, (2) flu shot is cancelled for a season, or (3) live in a home with slightly elevated radon. As a statistician, I would choose to have asymptomatic people interact as normal during the COVID-19 pandemic because it is safer than skipping the flu shot or living with radon – by far.

The field of statistics cannot determine what tradeoffs should be made when lives are at risk. It can only tell us the relative risks of various options. In the case of coronavirus lockdowns, we are not being consistent; we have never been this restrictive even when dealing with greater risks.

A reasonable critic could quibble with the assumptions that I have made about the risk of healthy persons engaging in public interactions during the pandemic.[102] If so, they should explain and defend their own assumptions. In the case of coronavirus lockdowns, there was no effort to directly quantify the risk of healthy individuals interacting in public as justification for the restrictions.

Even without hard numbers, officials should have known that fighting asymptomatic spread was not the key to the coronavirus battle. A March 12 article in the peer-reviewed journal *Eurosurveillance* wrote: "Currently, there is no clear evidence that COVID-19 asymptomatic persons can transmit SARS-CoV-2".[103] While evidence of asymptomatic spread has since emerged[104], studies estimate it accounts for roughly 10% of transmissions[105] – meaning a focus on properly isolating symptomatic cases would tackle 90% of the problem. [106]

The Tragedy of Closed Schools
The city of Vo in hard-hit Italy set a goal to test every member of their community for coronavirus – twice. In late February they succeeded in testing 86% of their population. In early March they tested 72%.

A single statement in their findings should indict America's indiscriminate approach: "No infections were detected in either survey in 234 tested children ranging from 0 to 10 years, despite some of them living in the same household as infected people."[107] For unknown reasons, COVID-19 does not generally endanger children. Children have contracted the disease, but rarely show symptoms let alone progression beyond a fever and cough.[108] They clear the disease quickly.

Among the many COVID-19 unknowns, the lack of imminent danger to children emerged early as an important, solid fact which has been confirmed by every COVID-19 study. The WHO Joint Mission report on China stated in February: "Of note, people interviewed by the Joint Mission Team could not recall episodes in which transmission occurred from a child to an adult."[109]

The decision to close schools has caused harm that does not need to be recounted here. The educational and social setbacks, particularly for those students who start out with disadvantages, will be felt for a long time. It was utterly unnecessary.

To withstand the pressure and maintain in-person education, or even just to re-open schools for May, would have required enormous political courage. It would have required the audacity of Sweden, which never closed primary schools. No Swedish children have died as a result.

9

WHERE WE GO FROM HERE

In this short book we have quantified and analyzed many aspects of the coronavirus pandemic and arrived at some useful conclusions. Yet, some aspects of the outbreak remain difficult to unravel. Why did Lombardy, Madrid, Brussels, New York, and New Orleans get hit so hard, while many major cities did not? This chapter is going to be a lot more speculative.

We know that New York and other hard-hit metropolises had some disadvantages in fighting COVID-19. Plagues have always hit urban centers the hardest because contact rates among citizens are so high (and hygiene levels are often lower, too).[110]

We have seen that individual-level responses keep disease growth from reaching exponential levels. But a necessary condition for these responses to kick in is knowledge that the disease is present. Several of the hardest-hit cities seem to have had introductions of the disease long before anyone knew the virus had exited China. I suspect these disadvantages are the primary culprits behind the most tragic situations.

Another factor in the wildly disparate COVID-19 effects may be different viral strains. Coronaviruses are known to be capable of rapid mutation,[111] and researchers in Washington state have been documenting the genetic relationship of COVID-19 strains based on differences in thousands of sequenced genomes.[112] Dr. Kari Stefansson, a researcher in Iceland, has speculated that mutations may be "responsible in some way, for how differently people respond to it."[113]

At this point we do not know what is causing such variations in COVID-19 experiences at the individual or geographic level, so it is very speculative to consider what path the coronavirus will take in the future. The warmth, humidity, and sunlight of summer may depress COVID-19 for a few months, as it does with other coronaviruses. When cold weather returns, the virus will have a full winter season rather than its partial first winter to spread, so there is the possibility it will come back strong.

But there are many reasons to think its spread will be constrained. Some individuals will have survived it once and may have immunity. The medical community has been learning how to better treat patients. And natural selection seems more likely to favor weaker strains, since killing a host is not in the best interest of the virus.[114] Evolutionary theorists typically predict a tradeoff where a virus mutates to an optimal virulence that is "below the maximum virulence level."[115]

In the short term, what should governments do? The same experts who told us to lockdown are telling us to stay locked down. Georgia's governor ended his "shelter-in-place" order effective April 30 and has allowed gyms and salons to re-open. The CovidActNow website predicts 98,000 deaths within 3 months as a result (using their model with an April 23 timestamp).[116]

A Georgia Tech professor has built a COVID simulator. When I compare the "current intervention" vs. the "minimal restrictions" scenarios, the Georgia deaths by August 31 go from 2,020 to 64,600.[117] A team from Harvard/MIT projects roughly 3,000 – 7,000 deaths depending on the level of activity that resumes in Georgia, while claiming deaths would only be 2,000 if the lockdown persisted.[118]

I have presented data and arguments to support my conclusion that states should largely re-open, tomorrow. Elementary schools should re-open, yesterday. The lockdowns have not prevented millions of deaths and ending them will not cause hundreds of thousands of deaths. Fortunately, from a scientific perspective, several states appear to be pressing towards "re-opening", so the second round of dire predictions can once again be tested.

Re-opening does not have to mean everything returns to normal. Nursing homes need to remain on lockdown for the foreseeable future, as their vulnerable residents have borne the brunt of this pandemic. The large gathering ban limits could be raised incrementally. Businesses and organizations will naturally be cautious; the government should trust them.

What should individuals do? I speak here as a statistician, not a clinician.

Since my basic argument is that the world is now not much riskier than it was in the past, the most important advice is the same as before COVID-19.

Wash your hands thoroughly and frequently. If you have a fever or cough isolate yourself from others. Do not interact closely with others exhibiting symptoms. Try to implement a lifestyle that prevents hypertension and Type II diabetes, which can be deadly when paired with COVID-19. Elderly and at-risk individuals should be especially careful.

Should we all be wearing masks? A systematic review of the effectiveness of facemasks concluded: "Further high quality trials are needed to assess when wearing a facemask in the community is most likely to be protective." After 31 separate studies of facemasks, the conclusion was that we need even further study.[119] The debate on facemasks will continue, but it is clear enough that facemasks in public offer weak protection, if any.

What should individuals and government do together? This is perhaps one of the most important questions we can ask, especially if the re-opening of the economy does not bring disastrous deaths.

The disruptive restrictions we have studied arrived via political executives. Governors closed businesses and forced individuals to "shelter-in-place" by declaration, not by legislation. There was no discussion or debate. No one played the role of "devil's advocate." These are ingredients that successful statistics and successful science demand.

When Neil Ferguson gave testimony to a parliamentary select committee, he clarified that Britain was on course for less than 20,000 deaths rather than his prior 550,000 estimate that had made headlines.[120] Even though he has since joined the chorus warning against re-opening, the atmosphere of a committee allowed for discussion and provided a venue for Mr. Ferguson to revise his estimate of a key input so that the public was aware.

To encourage a deliberative process, states should consider revising the emergency powers they give to their executives. Allowing an emergency order to stand for only a limited period of time without concurrence by a legislative chamber would both improve transparency and allow for broader input.

In the coming months, data on the COVID-19 epidemic will continue to emerge. Statisticians should continue the difficult task of making predictions but should also participate in studying results. I plan to follow the numbers.

I want to know.

APPENDIX

See below for additional supporting information from Chapter 6. Note that the source for timing of business closures and stay-at-home orders is the Institute for Health Metrics and Evaluation. States which never implemented a given measure were assigned a date of May 1, corresponding to the day death rates were measured.

Figure 5: Excess COVID-19 deaths by days until non-essential business closure

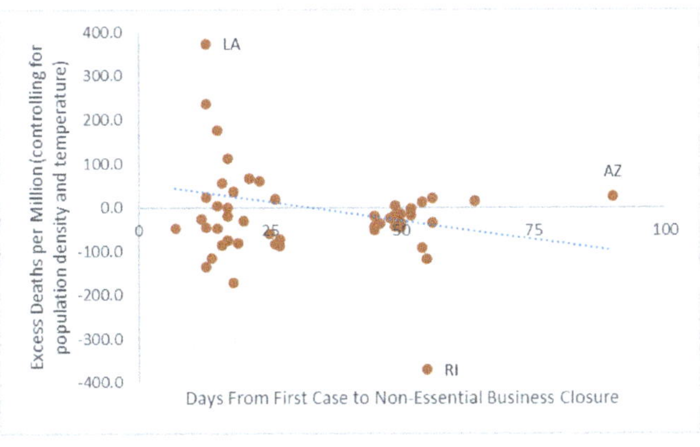

Table 3: Population data for the United States, as of 5/1/20

State	Deaths per Million	Population per Sq Km	Avg Temp (°F)	Median Age
Alabama	59	36.9	62.7	38.9
Alaska	12	0.5	32.0	34.0
Arizona	48	22.5	66.0	37.4
Arkansas	21	22.0	60.1	37.9
California	55	95.0	61.2	36.3
Colorado	148	19.6	46.3	36.6
Connecticut	653	286.7	48.6	40.8
Delaware	167	183.4	54.5	40.2
Florida	64	140.8	71.8	41.9
Georgia	113	67.1	62.3	36.5
Hawaii	11	84.4	73.0	38.9
Idaho	37	7.5	46.3	36.1
Illinois	192	89.6	51.4	37.9
Indiana	177	70.8	51.8	37.6
Iowa	54	21.4	48.1	38.1
Kansas	48	13.7	54.7	36.5
Kentucky	56	43.0	55.6	38.7
Louisiana	422	41.3	66.7	36.6
Maine	41	16.6	43.1	44.6
Maryland	208	235.8	54.6	38.6
Massachusetts	544	331.3	48.1	39.4
Michigan	388	67.6	46.6	39.7
Minnesota	67	26.3	43.0	37.9
Mississippi	94	24.6	63.5	37.2
Missouri	59	33.9	54.7	38.5
Montana	15	2.7	44.2	39.8
Nebraska	38	9.4	49.9	36.4
Nevada	87	9.8	57.3	37.9
New Hampshire	60	57.1	44.5	42.7
New Jersey	849	467.2	51.9	39.8
New Mexico	63	6.6	53.1	37.5
New York	1,227	161.0	48.2	38.7
North Carolina	41	78.2	58.7	38.6
North Dakota	31	4.0	40.8	35.1
Ohio	86	109.3	50.9	39.3
Oklahoma	59	21.7	60.1	36.4
Oregon	25	15.8	51.3	39.2
Pennsylvania	207	110.2	49.8	40.7
Rhode Island	264	392.7	49.3	39.9
South Carolina	52	61.3	61.7	39.2
South Dakota	24	4.3	45.5	36.8
Tennessee	31	60.8	58.1	38.7
Texas	30	39.1	66.0	34.4
Utah	15	13.6	49.6	30.7
Vermont	80	26.3	43.7	42.9
Virginia	69	80.8	55.5	38.1
Washington	114	40.5	50.5	37.6
West Virginia	26	29.8	53.1	42.4
Wisconsin	57	40.9	44.7	39.3
Wyoming	12	2.3	43.5	37.3

Table 4: Statewide emergency restrictions[121]

State	First Case Date	Initial Bus Ban Date	Non-Essential Bus Ban Date	Stay At Home Order Date	Time Until Init Bus Ban	Time Until NE Bus Ban	Time Until Stay At Home
Alabama	13-Mar	19-Mar	28-Mar	4-Apr	6	15	22
Alaska	12-Mar	17-Mar	28-Mar	28-Mar	5	16	16
Arizona	26-Jan	1-May	1-May	30-Mar	96	96	64
Arkansas	11-Mar	19-Mar	1-May	1-May	8	51	51
California	25-Jan	19-Mar	19-Mar	19-Mar	54	54	54
Colorado	5-Mar	17-Mar	26-Mar	26-Mar	12	21	21
Connecticut	8-Mar	16-Mar	23-Mar	1-May	8	15	54
Delaware	11-Mar	16-Mar	24-Mar	24-Mar	5	13	13
Florida	1-Mar	17-Mar	1-May	3-Apr	16	61	33
Georgia	2-Mar	24-Mar	1-May	3-Apr	22	60	32
Hawaii	6-Mar	17-Mar	25-Mar	25-Mar	11	19	19
Idaho	13-Mar	25-Mar	25-Mar	25-Mar	12	12	12
Illinois	25-Jan	16-Mar	21-Mar	21-Mar	51	56	56
Indiana	6-Mar	16-Mar	24-Mar	25-Mar	10	18	19
Iowa	8-Mar	17-Mar	1-May	1-May	9	54	54
Kansas	7-Mar	1-May	1-May	30-Mar	55	55	23
Kentucky	6-Mar	16-Mar	26-Mar	1-May	10	20	56
Louisiana	9-Mar	17-Mar	22-Mar	23-Mar	8	13	14
Maine	12-Mar	18-Mar	25-Mar	2-Apr	6	13	21
Maryland	5-Mar	16-Mar	23-Mar	30-Mar	11	18	25
Massachusetts	1-Feb	17-Mar	24-Mar	1-May	45	52	90
Michigan	10-Mar	16-Mar	23-Mar	24-Mar	6	13	14
Minnesota	6-Mar	17-Mar	1-May	27-Mar	11	56	21
Mississippi	11-Mar	24-Mar	3-Apr	3-Apr	13	23	23
Missouri	7-Mar	23-Mar	1-May	6-Apr	16	55	30
Montana	11-Mar	20-Mar	26-Mar	26-Mar	9	15	15
Nebraska	6-Mar	19-Mar	1-May	1-May	13	56	56
Nevada	5-Mar	18-Mar	21-Mar	31-Mar	13	16	26
New Hampshire	2-Mar	16-Mar	28-Mar	27-Mar	14	26	25
New Jersey	4-Mar	16-Mar	21-Mar	21-Mar	12	17	17
New Mexico	11-Mar	16-Mar	24-Mar	1-May	5	13	51
New York	1-Mar	16-Mar	22-Mar	22-Mar	15	21	21
North Carolina	3-Mar	17-Mar	30-Mar	30-Mar	14	27	27
North Dakota	11-Mar	20-Mar	1-May	1-May	9	51	51
Ohio	9-Mar	15-Mar	23-Mar	23-Mar	6	14	14
Oklahoma	6-Mar	1-Apr	1-Apr	1-May	26	26	56
Oregon	29-Feb	17-Mar	1-May	23-Mar	17	62	23
Pennsylvania	6-Mar	18-Mar	23-Mar	1-Apr	12	17	26
Rhode Island	1-Mar	17-Mar	1-May	28-Mar	16	61	27
South Carolina	7-Mar	18-Mar	1-May	7-Apr	11	55	31
South Dakota	10-Mar	1-May	1-May	1-May	52	52	52
Tennessee	5-Mar	23-Mar	1-Apr	2-Apr	18	27	28
Texas	4-Mar	21-Mar	1-May	2-Apr	17	58	29
Utah	6-Mar	19-Mar	1-May	1-May	13	56	56
Vermont	8-Mar	17-Mar	25-Mar	24-Mar	9	17	16
Virginia	7-Mar	17-Mar	24-Mar	30-Mar	10	17	23
Washington	21-Jan	16-Mar	25-Mar	23-Mar	55	64	62
West Virginia	17-Mar	18-Mar	24-Mar	25-Mar	1	7	8
Wisconsin	29-Feb	17-Mar	25-Mar	25-Mar	17	25	25
Wyoming	11-Mar	19-Mar	1-May	1-May	8	51	51

ABOUT THE AUTHOR

Benjamin Hall has a PhD in Statistics from the University of Kentucky, with a concentration in Biostatistics. He lives in Louisville, Ky.

The views expressed in this book are the author's alone and should not be construed to represent the views of any of his collaborators, reviewers, employers, or associates past and present in any way.

NOTES

[1] Deese, Kaelan. (2020). Louisville doctor arrested, charged with strangulation after fighting teens over social distancing. Retrieved April 26, 2020, from https://thehill.com/homenews/news/491830-louisville-doctor-arrested-charged-with-strangulation-after-fighting-teens-over

[2] Vogt, Dustin. (2020). Maryville Baptist Church temporary restraining order denied by federal judge. Retrieved April 26, 2020, from https://www.wave3.com/2020/04/19/maryville-baptist-church-temporary-restraining-order-denied-by-federal-judge/

[3] Moreno, J. Edward (2020). Government health agency official: Coronavirus 'isn't something the American public need to worry about'. Retrieved April 26, from https://thehill.com/homenews/sunday-talk-shows/479939-government-health-agency-official-corona-virus-isnt-something-the

[4] Budryk, Zack. (2020). Fauci: 'If it looks like you're overreacting, you're probably doing the right thing'. Retrieved April 26, 2020, from https://thehill.com/homenews/sunday-talk-shows/487639-fauci-if-it-looks-like-youre-overreacting-youre-probably-doing-the

[5] Gillespie, Nick. (2020). Richard Epstein cops to a 'stupid gaffe' in controversial coronavirus essay that caught Trump administration attention. Retrieved April 26, 2020, from https://reason.com/2020/03/24/richard-epstein-cops-to-a-stupid-gaffe-in-controversial-coronavirus-essay-that-caught-trump-admin-attention/

[6] LaSalle, Gretchen. (2020). Coronavirus Pandemic – YOUR actions matter. Retrieved April 26, 2020, from https://gretchenlasallemd.com/coronavirus-pandemic-your-actions-matter/

[7] Lovelace, Berkeley and Higgins-Dunn, Noah. (2020). WHO says coronavirus death rate is 3.4% globally, higher than previously thought. Retrieved April 26, 2020, from https://www.cnbc.com/2020/03/03/who-says-coronavirus-death-rate-is-3point4percent-globally-higher-than-previously-thought.html

[8] Wang, Chaolong, et al. (2020). Evolving epidemiology and impact of non-pharmaceutical interventions on the outbreak of coronavirus disease 2019 in Wuhan, China. Retrieved April 26, 2020, from https://www.medrxiv.org/content/10.1101/2020.03.03.20030593v1.full.pdf+html

[9] Smith, Kyle. (2020). The ventilator shortage that wasn't. Retrieved April 26, 2020, from https://www.yahoo.com/news/ventilator-shortage-wasn-t-212614058.html

[10] Specktor, Brandon. (2020). Coronavirus: What is 'flattening the curve,' and will it work? Retrieved April 26, 2020, from https://www.livescience.com/coronavirus-flatten-the-curve.html

[11] Office of the Governor, State of Ohio. (2020). Ohio bans mass gatherings of 100 or more. Retrieved Apr 26, 2020, from https://governor.ohio.gov/wps/portal/gov/governor/media/news-and-media/bans-mass-gatherings-of-100-or-more

[12] Boone, Kyle. (2020). 2020 NCAA Tournament games to be played without fans in attendance due to threat of coronavirus. Retrieved April 26, 2020, from https://www.cbssports.com/college-basketball/news/2020-ncaa-tournament-games-to-be-played-without-fans-in-attendance-due-to-threat-of-coronavirus/

[13] McCormack, John. (2020). The senator who saw the coronavirus coming. Retrieved April 26, 2020, from https://www.nationalreview.com/2020/03/the-senator-who-saw-the-coronavirus-coming/

[14] Winfield, Nicole. (2020). 'Not a wave, a tsunami.' Italy hospitals at virus limit. Retrieved April 26, 2020, from https://www.pbs.org/newshour/health/not-a-wave-a-tsunami-italy-hospitals-at-virus-limit

[15] Mackenzie, James. (2020). Italy's coronavirus toll surges as worries grow over hospitals. Retrieved April 26, 2020, from https://www.reuters.com/article/us-health-coronavirus-italy/worries-grow-over-northern-hospitals-as-italys-coronavirus-toll-grows-idUSKBN2120XN

[16] Ferguson, Neil M. et al. (2020). Impact of non-pharmaceutical interventions (NPIs) to reduce COVID-19 mortality and healthcare demand. Retrieved April 26, 2020, from https://www.imperial.ac.uk/media/imperial-college/medicine/sph/ide/gida-fellowships/Imperial-College-COVID19-NPI-modelling-16-03-2020.pdf

[17] Fink, Sheri. (2020). White House takes new line after dire report on death toll. Retrieved April 26, 2020, from https://www.nytimes.com/2020/03/16/us/coronavirus-fatality-rate-white-house.html

[18] Weiter, Taylor. (2020). Kentucky could have 13,000 deaths with poor social distancing, new projection says. Retrieved April 26, 2020, from https://www.whas11.com/article/news/health/coronavirus/kentucky-could-have-13000-deaths-with-poor-compliance-new-projection-says/417-06c9c1f3-1995-4f10-9351-030dd0954387

[19] Chavez, Stella M. (2020). Amid COVID-19 spread, Dallas county orders residents to shelter-in-place. Retrieved Apr 26, 2020, from https://www.keranews.org/post/amid-covid-19-spread-dallas-county-orders-residents-shelter-place

[20] Municipality of Anchorage, Office of the Mayor. (2020). Assessment of two COVID-19 models to guide community intervention policies in Anchorage and Alaska. Retrieved April 26, 2020, from https://www.uaa.alaska.edu/academics/college-of-health/college/news/_documents/2020-03-25-COVID19-models-review-FINAL-3-30-20.pdf

[21] Ewing, Philip. (2020). Coronavirus Task Force details 'sobering' data behind its extended guidelines. Retrieved April 26, 2020, from https://www.npr.org/2020/03/31/823916343/coronavirus-task-force-set-to-detail-the-data-that-led-to-extension-of-guideline

[22] Wise, Justin. (2020). Birx: US could see up to 200,000 coronavirus deaths if 'we do things almost perfectly'. Retrieved April 26, 2020, from https://thehill.com/homenews/administration/490138-birx-says-us-could-have-up-to-200000-coronavirus-deaths-if-we-do

[23] Frieden, Tom. (2020). Could coronavirus kill a million Americans? Retrieved April 26, 2020, from https://www.thinkglobalhealth.org/article/could-coronavirus-kill-million-americans

[24] Wilson, Reid. (2020). Worst-case coronavirus models show massive US toll. Retrieved April 26, 2020, from https://thehill.com/homenews/state-watch/487489-worst-case-coronavirus-models-show-massive-us-toll

[25] Pueyo, Tomas. (2020). Coronavirus Articles: Endorsements and public shares. Retrieved April 26, 2020, from https://medium.com/tomas-pueyo/coronavirus-articles-endorsements-fdc68614f8e3

[26] Johnson, Andrew and Bravo, Christina. (2020). Governor issues statewide 'stay-at-home' order; projects 56% of Californians could get coronavirus. Retrieved April 26, 2020, from https://www.nbcsandiego.com/news/local/governor-projects-56-of-californians-may-be-infected-with-coronavirus-in-next-8-weeks/2289281/

[27] Goudie, Chuck et al. (2020). With shortage of ventilators to treat COVID-19 patients, hospitals face ethical and logistical questions. Retrieved April 26, 2020, from https://abc7chicago.com/coronavirus-treatment-cases-symptoms/6050639/

[28] Yong, Ed. (2020). How the pandemic will end. Retrieved April 26, 2020, from https://www.theatlantic.com/health/archive/2020/03/how-will-coronavirus-end/608719/

[29] Ladapo, Joseph. (2020). Coronavirus pandemic: we were caught unprepared. It is too late for shutdowns to save us. Retrieved April 26, 2020, from https://www.usatoday.com/story/opinion/2020/03/24/coronavirus-shutdowns-worth-public-health-system-unprepared-column/2898324001/

[30] Inglesby, Tom. (2020). Coronavirus: no, we aren't even close to ready to ease up on social distancing. Retrieved April 26, 2020, from https://www.usatoday.com/story/opinion/2020/03/26/coronavirus-pandemic-growing-too-fast-stop-social-distancing-column/5083173002/

[31] WKYC Staff. (2020). Ohio Gov. Mike DeWine: Cleveland Clinic projections show hospitals will be hit hard in 2 weeks, may need triple capacity for mid-May coronavirus peak. Retrieved Apr 26, 2020 from https://www.wkyc.com/article/news/health/coronavirus/ohio-gov-mike-dewine-cleveland-clinic-projections-show-hospitals-will-be-hit-hard-in-2-weeks-peak-may-not-come-until-mid-may/95-46798fea-92c4-4e21-93c6-82a37ebbde2c

[32] Emanuel, Ezekiel. (2020). We can safely restart the economy in June. Here's how. Retrieved April 26, 2020, from https://www.nytimes.com/2020/03/28/opinion/coronavirus-economy.html

[33] Yong, Ed. (2020). How the pandemic will end. Retrieved April 26, 2020, from https://www.theatlantic.com/health/archive/2020/03/how-will-coronavirus-end/608719/

[34] Unless otherwise noted, COVID-19 deaths and rates are sourced from worldometers.info/coronavirus.

[35] Murphy, Brendan. (2020). COVID-19: States call on early medical school grads to bolster workforce. Retrieved April 26, 2020, from https://www.ama-assn.org/delivering-care/public-health/covid-19-states-call-early-medical-school-grads-bolster-workforce

[36] Earley, Neal. (2020). State cuts red tape for retired doctors, nurses who want to rejoin front lines to fight coronavirus. Retrieved April 26, 2020, from https://chicago.suntimes.com/politics/2020/3/23/21191544/state-retired-doctors-nurses-coronavirus-license-expire-renew

[37] Scott, Dylan. (2020). Hospital are laying off workers in the middle of the coronavirus pandemic. Retrieved April 26, 2020, from https://www.vox.com/2020/4/8/21213995/coronavirus-us-layoffs-furloughs-hospitals

[38] Sorace, Stephen. (2020). Washington's field hospital to be dismantled before treating a patient, sent to states worse-hit by coronavirus. Retrieved April 26, 2020, from https://www.foxnews.com/us/washington-field-hospital-coronavirus-dismantle

[39] Hernandez, Daniela. (2020). Coronavirus ravages the lungs. It also affects the brain. Retrieved May 2, 2020, from https://www.wsj.com/articles/coronavirus-ravages-the-lungs-it-also-affects-the-brain-11586896119

[40] CDC. (2020). Past pandemics. Retrieved May 2, 2020 from https://www.cdc.gov/flu/pandemic-resources/basics/past-pandemics.html

[41] CDC. (2020). 1918 pandemic influenza historic timeline. Retrieved May 2, 2020 from https://www.cdc.gov/flu/pandemic-resources/1918-commemoration/pandemic-timeline-1918.htm

[42] CDC. (2020). Table 1: Estimated influenza disease burden, by season – United States, 2010-11 through 2018-19 influenza seasons. Retrieved April 26, 2020, from https://www.cdc.gov/flu/about/burden/past-seasons.html

[43] Faust, Jeremy Samuel. (2020). Comparing COVID-19 deaths to flu deaths is like comparing apples to oranges. Retrieved May 2, 2020 from https://blogs.scientificamerican.com/observations/comparing-covid-19-deaths-to-flu-deaths-is-like-comparing-apples-to-oranges/

[44] UW Health. (2020). Is it flu or an influenza-like illness? Retrieved May 2, 2020 from https://www.uwhealth.org/health-wellness/is-it-the-influenza-or-an-influenza-like-illness/51692

[45] IHME. (2020). COVID-19 Projections. Retrieved May 2, 2020, from https://covid19.healthdata.org/united-states-of-america

[46] Tegnell, Anders. (2007). The impact of pandemics on society. Retrieved April 26, 2020, from https://kkrva.se/wp-content/uploads/Artiklar/075/kkrvaht_5_2007_6.pdf

[47] Sydsvenskan. (2019). Hundratals unga fick narkolepsi av vaccin. Retrieved April 26, 2020, from

https://www.sydsvenskan.se/2019-03-24/vaccinet-som-forstorde-livet-for-hundratals

[48] BBC Staff. (2020). Coronavirus: Lombardy region announces stricter measures. Retrieved April 26, 2020, from https://www.bbc.com/news/world-europe-51991972

[49] Elliott, Christopher. (2020). France's coronavirus lockdown offers a preview of restrictions we may see in American. Retrieved Apr 26, 2020, from https://www.usatoday.com/story/travel/news/2020/04/02/coronavirus-frances-lockdown-offers-preview-what-may-happen-us/5103335002/

[50] The Local Staff. (2020). After lockdown: are Denmark and Norway's restrictions now like Sweden's? Retrieved April 26, 2020, from https://www.thelocal.no/20200421/explained-are-denmark-and-norways-restrictions-still-tougher-than-swedens

[51] Baker, Sinead. (2020). Denmark rushed to lock down before almost every other country. Now its response is so far ahead that it's starting to remove restrictions. Retrieved Apr 26, 2020 from https://www.businessinsider.com/coronavirus-how-denmark-reached-stage-of-easing-lockdown-restrictions-2020-4

[52] Financial Times Staff. (2020). In Sweden's far north, the skiing continues. Retrieved April 14, 2020, from https://www.ft.com/content/93c78a76-7b0a-11ea-bd25-7fd923850377

[53] Ellyatt, Holly. (2020). No lockdown here: Sweden defends its more relaxed coronavirus strategy. Retrieved April 26, 2020, from https://www.cnbc.com/2020/03/30/sweden-coronavirus-approach-is-very-different-from-the-rest-of-europe.html

[54] Lister, Tim and Shukla, Sebastian. (2020). Sweden challenges Trump – by refusing to lock down. Retrieved April 26, 2020, from https://www.cnn.com/2020/04/10/europe/sweden-lockdown-turmp-intl/index.html

[55] Erixon, Fredrik. (2020). No lockdown, please, we're Swedish. Retrieved April 26, 2020, from https://spectator.us/lockdown-please-swedish/

[56] Tahir, Tariq. (2020). 'Horror movie' coronavirus will infect half of Sweden's population in just four weeks as nation refuses to lock down, scientist claims. Retrieved April 26, 2020, from https://www.the-sun.com/news/626149/coronavirus-infect-half-swedens-population/

[57] Savage, James. (2020). Why Sweden is Europe's coronavirus outlier and what it means. Retrieved April 26, 2020, from https://www.newstatesman.com/world/europe/2020/04/why-sweden-europe-s-coronavirus-outlier-and-what-it-means

[58] Bloomberg Staff. (2020). Did Sweden get its coronavirus strategy horribly wrong? Retrieved Apr 26, 2020 from https://www.scmp.com/news/world/europe/article/3078587/did-sweden-get-its-coronavirus-strategy-horribly-wrong

[59] Masters, Kate. (2020). Northam sticks with decision not to close bars, restaurants and other businesses. Retrieved April 26, 2020, from https://www.virginiamercury.com/blog-va/northam-sticks-with-decision-not-to-close-bars-restaurants-and-other-businesses/

[60] McCarthy, Tom. (2020). Disunited states of America: responses to coronavirus shaped by hyper-partisan politics. Retrieved April 26, 2020, from https://www.theguardian.com/us-news/2020/mar/29/america-states-coronavirus-red-blue-different-approaches

[61] Koopman, Christopher and Philpot, Anne. (2016). The state of certificate-of-need laws in 2016. Retrieved April 26, 2020, from https://www.mercatus.org/publications/corporate-welfare/state-certificate-need-laws-2016

[62] Boden, Anastasia and Philpot, Anne. (2020). Certificate of need laws limited healthcare capacities in the years leading up to COVID-19. Retrieved April 26, 2020, from https://pacificlegal.org/certificate-of-need-laws-healthcare-capacity-covid-19/

[63] Mahoney, Bill. (2020). Here's every law and regulation Cuomo had suspended during coronavirus crisis. Retrieved April 26, 2020, from https://www.politico.com/states/new-york/albany/story/2020/03/19/every-law-and-regulation-suspended-by-cuomo-during-the-coronavirus-crisis-1268180

[64] Centers for Medicare & Medicaid Services Newsroom. (2020). Retrieved April 26, 2020, from https://www.cms.gov/newsroom/fact-sheets/medicare-telemedicine-health-care-provider-fact-sheet

[65] Bluth, Rachel. (2020). California resists push to lift limits on nurse practitioners during Covid-19 pandemic. Retrieved April 26, 2020, from https://www.statnews.com/2020/04/17/california-resists-push-to-lift-limits-on-nurse-practitioners-during-covid-19-pandemic/

[66] Charlton, John and Ash, Andrea. (2020). Two new laws are changing the way Kentuckians get alcohol. Retrieved April 26, 2020, from https://www.whas11.com/article/news/investigations/focus/kentucky-laws-alcohol-coronavirus/417-2a920b53-1c51-434b-990f-590234dc7c55

[67] Fleming, Deirdre. (2020). Fishing is free in Main through April. Retrieved April 26, 2020, from https://www.pressherald.com/2020/03/20/fishing-is-free-in-maine-through-april/

[68] Peterson, Hayley. (2020). Walmart clarifies its policy on the sale of nonessential goods after a shopper said she was barred from buying a baby car seat. Retrieved on April 26, 2020, from https://www.businessinsider.com/walmart-clarifies-policy-on-nonessential-items-after-car-seat-backlash-2020-4

[69] Starnes, Todd. (2020). Louisville mayor forbids churches from holding Easter Sunday drive-in services. Retrieved April 26, 2020, from https://www.toddstarnes.com/faith/louisville-mayor-forbids-churches-from-holding-easter-sunday-drive-in-services/

[70] Shiver, Phil. (2020). The city dumped 37 tons of sand on their skatepark as part of COVID lockdown, so skaters used it to ride dirt bikes. Then they dug up the sand and continued skating. Retrieved April 26, 2020, from

https://www.theblaze.com/news/san-clemente-skatepark-dirt-bikes

[71] Puhak, Janine. (2020). Washington state anglers protest fishing ban amid coronavirus outbreak: 'Let us fish'. Retrieved Apr 26, 2020, from https://www.foxnews.com/great-outdoors/coroanvirus-washington-state-anglers-fishing-ban-protest

[72] Ebrahimji, Alisha. (2020). Pennsylvanians are driving out of state to buy liquor, so neighboring states are cracking down. Retrieved May 2, 2020 from https://www.cnn.com/2020/04/15/us/pennsylvania-liquor-sales-ohio-west-virginia-trnd/index.html

[73] Ianllonardo, Marisa. (2020). How two US cities responded to the 1918 flu pandemic very differently – and what we can learn from those mistakes. Retrieved April 20, 2020 from https://www.businessinsider.com/history-of-how-st-louis-vs-philadelphia-treated-1918-flu-pandemic-2020-4

[74] Hatchett, Richard. (2007). Public health interventions and epidemic intensity during the 1918 influenza pandemic. Retrieved May 2, 2020 from https://www.pnas.org/content/104/18/7582

[75] Coren, Michael. (2020). A tale of two cities: why social distancing works. Retrieved May 2, 2020 from https://www.christiancentury.org/article/news/tale-two-cities-why-social-distancing-works

[76] Barry, John. (2017). How the horrific 1918 flu spread across America. Retrieved May 2, 2020 from https://www.smithsonianmag.com/history/journal-plague-year-180965222/

[77] Markel, Howard. (2007). Nonpharmaceutical interventions implemented by US cities during the 1918-1919 influenza pandemic. *JAMA*. Retrieved May 2, 2020 from https://jamanetwork.com/journals/jama/fullarticle/208354?appId=scweb

[78] Courtemanche, Charles et al. (2020). Did social distancing measures in Kentucky help to flatten the COVID-19 curve? Retrieved May 2, 2020 from http://isfe.uky.edu/sites/ISFE/files/research-pdfs/NEWISFE%20Standardized%20Cover%20Page%20-%20Did%20Social%20Distancing%20Measures%20in%20Kentucky.pdf

[79] Ghebreyesus, Tedros. (2020). WHO director-general's opening remarks at the media briefing on COVID-19 – 3 March 2020. Retrieved April 26, 2020, from https://www.who.int/dg/speeches/detail/who-director-general-s-opening-remarks-at-the-media-briefing-on-covid-19---3-march-2020

[80] Carmeli, Oded. (2020). Trump is right about the coronavirus. The WHO is wrong,' says Israeli expert. Retrieved April 26, 2020, from https://www.haaretz.com/israel-news/.premium.MAGAZINE-israeli-expert-trump-is-right-about-covid-19-who-is-wrong-1.8691031

[81] Courtemache, Charles et al. (2020). Did social-distancing measures in Kentucky help to flatten the COVID-19 curve? Retrieved May 2, 2020 from http://isfe.uky.edu/sites/ISFE/files/research-pdfs/NEWISFE%20Standardized%20Cover%20Page%20-%20Did%20Social%20Distancing%20Measures%20in%20Kentucky.pdf

[82] Johnson, Carla K. (2020). Scientist links 2 state outbreaks with genetic fingerprints. Retrieved Apr 26, 2020 from https://abcnews.go.com/Technology/wireStory/scientist-links-state-outbreaks-genetic-fingerprints-69540345

[83] Mallapaty, Smriti. (2020). What the cruise-ship outbreaks reveal about COVID-19. Retrieved April 26, 2020, from https://www.nature.com/articles/d41586-020-00885-w

[84] Russell, Timothy W. (2020). Estimating the infection and case fatality ratio for COVID-19 using age-adjusted data from the outbreak on the Diamond Princess cruise ship. Retrieved April 26, 2020, from https://www.medrxiv.org/content/10.1101/2020.03.05.20031773v2.full.pdf+html

[85] Bendavid, Eran et al. (2020). COVID-19 antibody seroprevalence in Santa Clara County, California. Retrieved April 26, 2020, from https://www.medrxiv.org/content/10.1101/2020.04.14.20062463v1.full.pdf+html

[86] Sutton, Desmond et al. (2020). Universal screening for SARS-CoV-2 in women admitted for delivery. Retrieved April 26, 2020 from https://www.nejm.org/doi/full/10.1056/NEJMc2009316?query=featured_home

[87] Columbia University Irving Medical Center. (2020). In New York City, 1 in 7 expectant mothers test positive for coronavirus. Retrieved April 27, 2020 from https://www.cuimc.columbia.edu/news/new-york-city-1-7-expectant-mothers-test-positive-coronavirus

[88] Chappell, Bill. (2020). 73% of inmates at an Ohio prison test positive for coronavirus. Retrieved April 27, 2020 from https://www.npr.org/sections/coronavirus-live-updates/2020/04/20/838943211/73-of-inmates-at-an-ohio-prison-test-positive-for-coronavirus

[89] Ohio Department of Rehabilitation and Correction. (2020). COVID-19 inmate testing. Retrieved May 2, 2020 from https://drc.ohio.gov/Portals/0/DRC%20COVID-19%20Information%2004-19-2020%20%201305.pdf

[90] Government of Iceland. (2020). Large scale testing of general population in Iceland underway. Retrieved April 27, 2020 from https://www.government.is/news/article/2020/03/15/Large-scale-testing-of-general-population-in-Iceland-underway/

[91] CDC COVID-19 Response Team. (2020). Severe outcomes among patients with coronavirus disease 2019 (COVID-19) – Unites States, February 12 – March 16, 2020. Retrieved April 27, 2020 from https://www.cdc.gov/mmwr/volumes/69/wr/mm6912e2.htm?s_cid=mm6912e2_w

[92] Zimmerman, Petra and Nigel, Curtis. (2020). Coronavirus infections in children including COVID-19. Retrieved Apr 27, 2020 from https://journals.lww.com/pidj/Fulltext/2020/05000/Coronavirus_Infections_in_Children_Including.1.aspx

[93] Henderson, Max et al. (2020). Retrieved April 28, 2020 from https://covidactnow.org/

[94] Stevens, Harry. (2020). Why outbreaks like coronavirus spread exponentially, and how to "flatten the curve".

Retrieved April 28, 2020 from https://www.washingtonpost.com/graphics/2020/world/corona-simulator/

[95] Chowell, Gerardo et al. (2016). Mathematical models to characterize early epidemic growth: a review. *Phys Life Rev, 18*, pp. 66-97, https://www.ncbi.nlm.nih.gov/pmc/articles/PMC5348083/

[96] Viboud, Cecile et al. (2016). A generalized-growth model to characterize the early ascending phase of infectious disease outbreaks. *Epidemics (15)*, pp. 27-37, https://www.ncbi.nlm.nih.gov/pmc/articles/PMC4903879/

[97] Yglesias, Matthew. (2020). Opening up the economy won't save the economy. Retrieved April 28, 2020 from https://www.vox.com/2020/4/22/21228651/opening-up-save-economy-trump-coronavirus-pandemic-shutdown

[98] Boyd, Milo. (2020). Sweden warned thousands will die after 'playing Russian roulette' with public. Retrieved Apr 28, 2020 from https://www.mirror.co.uk/news/world-news/sweden-warned-prepare-thousands-deaths-21819683

[99] CDC. (2020). Radon: we track that! Retrieved May 2, 2020 from https://www.cdc.gov/nceh/features/trackingnetwork/index.html

[100] EPA. (2020). EPA assessment of risks from radon in homes. Retrieved May 2, 2020 from https://www.epa.gov/sites/production/files/2015-05/documents/402-r-03-003.pdf

[101] EPA. (2020). The national radon action plan – a strategy for saving lives. Retrieved May 2, 2020 from https://www.epa.gov/radon/national-radon-action-plan-strategy-saving-lives

[102] An attempt to inflate the risk of asymptomatic COVID-19 spread is likely to rely upon assumptions of exponential growth and high death rate, which we have already refuted

[103] Mizumoto et al. (2020). Estimating the asymptomatic proportion of coronavirus disease 2019 (COVID-19) cases on board the Diamond Princess cruise ship, Yokohama, Japan, 2020. Retrieved April 28, 2020 from https://www.eurosurveillance.org/content/10.2807/1560-7917.ES.2020.25.10.2000180

[104] Bai, Yan et al. (2020). Presumed asymptomatic carrier transmission of COVID-19. Retrieved April 28, 2020 from https://jamanetwork.com/journals/jama/fullarticle/2762028

[105] Wei, Wycliffe E. et al. (2020). Presymptomatic transmission of SARS-CoV-2 – Singapore, January 23-March 16, 2020. Retrieved April 28, 2020 from https://www.cdc.gov/mmwr/volumes/69/wr/mm6914e1.htm

[106] Du, Zhanwei et al. (2020). Serial interval of COVID-19 among publicly reported confirmed cases. Retrieved April 28, 2020 from https://wwwnc.cdc.gov/eid/article/26/6/20-0357_article

[107] Lavezzo, Enrico et al. (2020). Suppression of COVID-19 outbreak in the municipality of Vo, Italy. Retrieved April 26, 2020, from https://www.medrxiv.org/content/10.1101/2020.04.17.20053157v1.full.pdf+html

[108] Jiehao, Cai et al. (2020). A case series of children with 2019 novel coronavirus infection: clinical and epidemiological features. Retrieved April 26, 2020 from https://academic.oup.com/cid/advance-article/doi/10.1093/cid/ciaa198/5766430

[109] World Health Organization. (2020). Report of the WHO-China Joint Mission on Coronavirus Disease 2019 (COVID-19). Retrieved April 26, 2020 from https://www.who.int/docs/default-source/coronaviruse/who-china-joint-mission-on-covid-19-final-report.pdf

[110] Gould, S.E. (2011). Plague and the city. Retrieved April 28, 2020 from https://blogs.scientificamerican.com/lab-rat/plague-and-the-city/

[111] Zimmerman, Petra and Curtis, Nigel. (2020). An overview of the epidemiology, clinical features, diagnosis, treatment and prevention options in children. Retrieved April 28, 2020 from https://journals.lww.com/pidj/Fulltext/2020/05000/Coronavirus_Infections_in_Children_Including.1.aspx

[112] NextStrain.org

[113] John, Tara. (2020). Coronavirus: Iceland testing finds 50% of cases have no symptoms. Retrieved Apr 28, 2020 from https://www.mercurynews.com/2020/04/01/coronavirus-iceland-testing-finds-50-of-cases-have-no-symptoms/

[114] Payne, Rebecca et al. (2014). Impact of HLA-driven HIV adaptation on virulence in populations of high HIV seroprevalence. Retrieved April 28, 2020 from https://www.pnas.org/content/early/2014/11/26/1413339111

[115] Jensen, Knut Helge et al. (2006). Empirical support for optimal virulence in a castrating parasite. Retrieved April 28, 2020 from https://www.ncbi.nlm.nih.gov/pmc/articles/PMC1470460/

[116] CovidActNow.org, retrieved April 28, 2020 from https://covidactnow.org/us/ga

[117] COVID-19 Simulator. (2020). Retrieved April 28, 2020.

[118] Bredderman, William and Messer, Olivia. (2020). Georgia, Florida, and Mississippi are veering toward a terrifying premature end to their COVID-19 lockdowns, according to a new pandemic analysis. Retrieved April 28, 2020 from https://www.thedailybeast.com/ending-coronavirus-lockdowns-in-mississippi-georgia-and-florida-could-doom-thousands

[119] Brainard, Julii Suzanne. (2020) Facemasks and similar barriers to prevent respiratory illness such as COVID-19: a rapid systematic review. Retrieved April 28, 2020 from https://www.medrxiv.org/content/10.1101/2020.04.01.20049528v1

[120] Adam, David. (2020). UK has enough intensive care units for coronavirus, expert predicts. Retrieved Apr 28, 2020 from https://www.newscientist.com/article/2238578-uk-has-enough-intensive-care-units-for-coronavirus-expert-predicts/

[121] IHME. (2020). Retrieved on May 2, 2020, from https://covid19.healthdata.org/united-states-of-america